A Curious Book of Scales & Units of Measurement

I0409502

Compiled by
Zeeshan Mahmud

Welcome to a world where knowing these scales could be worth in gold! This reference booklet is a unique collection of a curious bibliophile and a delightful journey through the wild and wacky world of quantification.

Ever wondered why a chili pepper can make you sweat bullets or how we rate the intensity of earthquakes? Buckle up, because this book is your whimsical guide to a hundred scales that make you go, "Wow, I didn't know that was a thing!"

From the sizzle of the Scoville Scale to the seismic swagger of the Richter Scale, you're about to explore a universe where numbers and measurements get downright quirky. Forget complicated equations and dull textbooks; here, you'll learn about the tools we humans use to measure and classify our world in the most colorful, creative, and lighthearted way possible.

So, whether you're a math maverick, a science enthusiast, or just someone who loves discovering the "huh, I never knew that" moments, this manual promises to be your zaniest ride through the world of quantification. It's the book where you'll find the unexpected, the amusing, and the downright bizarre in the world of scales. Get ready to laugh, learn, and lighten up with scales like you've never seen them before!

A Brief History of Measurement Through Time

The ancient Egyptians wielded the cubit as their measure of choice. This remarkably intuitive unit, roughly equivalent to the length of a forearm, found its place in the construction of awe-inspiring pyramids that have withstood the test of millennia. From Egypt, we venture to ancient Greece, where the likes of Euclid established the foundations of geometric precision, laying the groundwork for measurement systems that would span centuries.

The Silk Road emerges as a pivotal chapter in our narrative—a bustling highway of ideas and goods connecting East and West. Alongside spices and silks, it facilitated the exchange of innovative measurement techniques. Merchants haggled over the weight of exotic treasures, while philosophers like Hipparchus delved into the nuances of angular measurement. These units transcended mere numbers; they were reflections of cultural identities and symbols of human ingenuity.

The Renaissance era ushered in an age of profound transformation. Galileo Galilei, Johannes Kepler, and Tycho Brahe emerged as luminaries, challenging age-old beliefs about our place in the universe. Pioneering instruments like astrolabes and quadrants guided the intrepid explorers of the time, from Columbus setting sail on uncharted seas to Magellan circumnavigating the globe. This was an era where precision became paramount.

The Victorian era marked the zenith of this relentless pursuit of precision. A fascination with accuracy led to the creation of the metric system, a universal language of measurement that transcended borders. Concurrently, the industrial revolution transformed manufacturing and trade, necessitating standardized measurements for everything from screws to steam engines.

In the 20th and 21st centuries, our quest for precision evolved in remarkable ways. Quantum mechanics and atomic clocks provided previously unimaginable accuracy, revolutionizing fields like navigation and telecommunications. The advent of space exploration demanded measurements capable of reaching the furthest reaches of our universe. Today, we stand on the precipice of discovery, with cutting-edge technology pushing the boundaries of what we can measure and comprehend.

Scales of Measurement

1. **Ringelmann Scale**: Measures the opacity or density of smoke.

2. **Beaufort Scale**: Rates wind speed at sea.

3. **Mohs Scale of Mineral Hardness**: Rates minerals based on scratch resistance.

4. **Fahrenheit Scale**: Measures temperature in the United States.

5. **Celsius Scale**: Measures temperature in most of the world.

6. **Kelvin Scale**: Measures temperature in the scientific community.

7. **pH Scale**: Measures the acidity or alkalinity of a solution.

8. **Richter Scale**: Measures the magnitude of earthquakes.

9. **Decibel Scale**: Measures the intensity of sound.

10. **Scoville Scale**: Measures the spiciness or heat of chili peppers.

11. **Hertzsprung-Russell Diagram**: Classifies stars based on luminosity and temperature.

12. **Brix Scale**: Measures sugar content in liquids, used in winemaking.

13. **Luminance Scale**: Measures the brightness of light.

14. **Brinell Hardness Scale**: Measures hardness of materials through indentation.

15. **Likert Scale**: Measures attitudes and opinions in surveys.

16. **Gini Coefficient**: Measures income inequality in a population.

17. **Bodily-Kinesthetic Intelligence Scale**: Part of Howard Gardner's theory of multiple intelligences.

18. **Modified Ashworth Scale**: Measures muscle spasticity in medical assessments.

19. **Modified Rankin Scale**: Measures disability or dependence after a stroke.

20. **Glasgow Coma Scale**: Assesses level of consciousness in patients with brain injuries.

21. **Karnofsky Performance Scale**: Evaluates functional status of cancer patients.

22. **Newton's Laws of Motion**: Describe relationship between body and forces acting upon it.

23. **Hounsfield Scale**: Measures radiodensity in medical imaging, used in CT scans.

24. **Apgar Score**: Rates physical condition of newborns immediately after birth.

25. **Borg Rating of Perceived Exertion Scale**: Measures perceived exertion during physical activity.

26. **Pain Scale (e.g., Visual Analog Scale)**: Quantifies intensity of pain experienced by a patient.

27. **Glasgow Outcome Scale**: Assesses overall outcome of patients with brain injuries.

28. **Epworth Sleepiness Scale**: Measures daytime sleepiness and likelihood of dozing off.

29. **Maslow's Hierarchy of Needs**: Classifies human needs in a hierarchical structure.

30. **Carnegie Classification of Institutions of Higher Education**: Classifies U.S. colleges and universities.

31. **Turing Test**: Evaluates machine's ability to exhibit intelligent behavior.

32. **Berkson's Paradox**: Relates to probability of conditional events.

33. **Economic Freedom Index**: Measures degree of economic freedom in countries.

34. **Hazardous Materials Identification System (HMIS)**: Labels hazardous materials based on properties.

35. **Peltier Scale**: Rates quality of diamonds based on color.

36. **Modified Newtonian Dynamics (MOND)**: An alternative theory of gravity.

37. **Scoville Heat Scale for Peppers**: Rates spiciness of chili peppers.

38. **Bradford Scale**: Used in workforce management to monitor employee absenteeism.

39. **Happiness Index (e.g., World Happiness Report)**: Ranks countries by citizens' subjective well-being.

40. **Kardashev Scale**: Measures civilization's level of technological advancement.

41. **Geiger-Nuttall Law**: Relates decay constant of radioactive substance to decay mode.

42. **Morse Scale**: Rates the pain and discomfort of kidney stones.

43. **Arithmetic Mean Scale**: Measures central tendency in statistics.

44. **Saffir-Simpson Hurricane Wind Scale**: Rates hurricane intensity.

45. **Sedgwick Rafter Cellulose Scale**: Rates the flammability of materials.

46. **Michelin Star Rating**: Rates the quality of restaurants.

47. **Infectious Disease Severity Scale (e.g., CDC Severity Index)**: Rates the seriousness of infectious diseases.

48. **Kopp-Etchells Effectiveness Index**: Measures military unit effectiveness.

49. **International Nuclear Event Scale (INES)**: Rates nuclear incidents.

50. **Richmond Agitation-Sedation Scale**: Measures level of sedation in patients.

51. **Arthralgia Severity Scale**: Rates severity of joint pain.

52. **Katz Index of Independence in Activities of Daily Living**: Rates a person's ability to perform daily activities.

53. **Stress Scale (e.g., Holmes and Rahe Stress Scale)**: Measures life stressors.

54. **Achenbach System of Empirically Based Assessment (ASEBA)**: Rates child and adolescent psychopathology.

55. **Child-Pugh Score**: Measures liver disease severity.

56. **Mini-Mental State Examination (MMSE)**: Rates cognitive impairment.

57. **New York Heart Association (NYHA) Functional Classification**: Rates heart disease severity.

58. **Modified Hachinski Ischemic Score**: Rates likelihood of vascular dementia.

59. **Bishop Score**: Measures readiness for labor in pregnant women.

60. **Motor and Social Development Scale (e.g., Denver Developmental Screening Test)**: Rates child development.

61. **Scale of Evil (Hare Psychopathy Checklist)**: Rates psychopathic traits.

62. **Pepper Scale (e.g., Ghost Pepper, Carolina Reaper)**: Rates spiciness of chili peppers.

63. **Kessler Psychological Distress Scale (K10)**: Rates psychological distress in individuals.

64. **Fluoride Toothpaste Strength Scale (e.g., 1,000 ppm)**: Rates fluoride content.

65. **Allen Cognitive Level Screen**: Rates cognitive abilities of individuals with disabilities.

66. **Bartel Index**: Rates functional independence in patients.

67. **Von Frey Filament Test**: Measures sensitivity to touch in neuropathic pain.

68. **AIS (Abbreviated Injury Scale)**: Rates injury severity in trauma patients.

69. **Renal Risk Score**: Predicts risk of acute kidney injury.

70. **Glasgow-Blatchford Bleeding Score**: Rates upper gastrointestinal bleeding risk.

71. **DASH (Disabilities of the Arm, Shoulder, and Hand) Score**: Rates upper extremity disability.

72. **MEWS (Modified Early Warning Score)**: Rates patient deterioration.

73. **Zung Self-Rating Depression Scale**: Rates depressive symptoms.

74. **Hamilton Rating Scale for Depression**: Rates severity of depression.

75. **Rosenberg Self-Esteem Scale**: Rates self-esteem.

76. **PSQI (Pittsburgh Sleep Quality Index)**: Rates sleep quality.

77. **Ramsay Sedation Scale**: Rates level of sedation in patients.

78. **Global Assessment of Functioning (GAF)**: Rates psychological, social, and occupational functioning.

79. **MASCC (Multinational Association of Supportive Care in Cancer) Risk Index Score**: Rates risk of febrile neutropenia in cancer patients.

80. **Myers-Briggs Type Indicator (MBTI)**: Rates personality type.

81. **MIDAS (Migraine Disability Assessment)**: Rates migraine-related disability.

82. **Fugl-Meyer Assessment**: Rates motor function after stroke.

83. **Morse Fall Scale**: Rates fall risk in patients.

84. **Norton Scale**: Rates risk of pressure ulcers.

85. **Modified Rankin Scale for Stroke**: Rates disability after stroke.

86. **Hamilton Anxiety Rating Scale**: Rates severity of anxiety.

87. **Richmond-Turner Depression Scales**: Rates depressive symptoms.

88. **Bristol Stool Scale**: Rates stool consistency.

89. **Borg Dyspnea Scale**: Rates breathlessness.

90. **CAGE Questionnaire**: Rates alcohol dependency.

91. **Neurological Deficit Scale (e.g., NIH Stroke Scale)**: Rates neurological impairment.

92. **Zarit Burden Interview**: Rates caregiver burden.

93. **Lund-Mackay Sinusitis Staging**: Rates severity of sinusitis.

94. **Asepsis Score**: Rates surgical site infection risk.

95. **BODE Index**: Rates severity of chronic obstructive pulmonary disease (COPD).

96. **Hamilton Depression Rating Scale**: Rates severity of depression.

97. **Disease Activity Score (DAS28)**: Rates disease activity in rheumatoid arthritis.

98. **Kurtosis Scale**: Measures the distribution of data in statistics.

99. **Henmon-Nelson Tests of Mental Ability**: Rates cognitive abilities in children.

100. **Oxford Happiness Questionnaire**: Rates happiness and life satisfaction.

Units of Measurement

These bizarre units of measurement add a touch of humor and uniqueness to various fields of study and everyday life.

1. **Smoot**: Equal to Oliver R. Smoot's height, used to measure the Harvard Bridge in Boston.

2. **Hogshead**: A unit of volume for liquids, typically wine or whiskey, equal to 63 gallons.

3. **Furlong**: Primarily used in horse racing, equal to 1/8th of a mile or 220 yards.

4. **Banana Equivalent Dose (BED)**: Compares radiation exposure to the radiation emitted by a banana.

5. **Jiffy**: An informal unit of time, denoting a very short period, often around 1/100th or 1/60th of a second.

6. **Sverdrup**: Used to describe ocean currents, equal to one million cubic meters per second.

7. **Donkeypower**: Measures power, approximately 250 watts, inspired by the work capacity of a donkey.

8. **Shake**: A unit of time equal to 10 nanoseconds, used in nuclear physics.

9. **Warhol**: Measures fame, inspired by Andy Warhol's statement about everyone being world-famous for 15 minutes.

10. **Micromort**: Measures risk, representing a one-in-a-million chance of death.

11. **Sheppey**: A unit of time equal to 7 years, based on the idea of recurring life events.

12. **Barn**: Measures area in nuclear physics, representing a cross-sectional area for nuclear reactions.

13. **Jansky**: Measures the intensity of radio waves from celestial objects in radio astronomy.

14. **Beezer**: A humorous unit of quantity when a specific number is unknown or irrelevant.

15. **Smoosh**: A playful unit of pressure used to describe the sensation of a hug.

16. **QALY (Quality-Adjusted Life Year)**: Measures the value of medical interventions in terms of added years of healthy life.

17. **Waffle Iron**: A whimsical unit of area to describe the size of a waffle iron.

18. **Helen**: Measures beauty, humorously referenced in discussions about the "face that launched a thousand ships."

19. **Quirk**: Measures angular acceleration in particle physics, equivalent to 1 million billion radians per second squared.

20. **Siriometer**: A unit of astronomical distance, equal to the distance that light travels in one year.

21. **Cubit**: An ancient unit of length, typically based on the length of a forearm, used in various ancient civilizations.

22. **Romer**: An old unit of length used in Scandinavia, equal to approximately 1.067 meters.

23. **Ogdoas**: An ancient Egyptian unit of measurement for land area.

24. **Quadrans**: An ancient Roman coin worth one-quarter of an as.

25. **Drachma**: An ancient Greek unit of currency and weight, used for silver coins.

26. **Lug**: A unit of volume used for alcoholic beverages in Ireland, equal to about 134 liters.

27. **Chestnut**: A unit of time equal to 24 days.

28. **Pood**: A Russian unit of weight, equal to 16.38 kilograms.

29. **Koku**: A Japanese unit of volume for rice, equal to about 180.4 liters.

30. **Scruple**: An apothecary unit of weight, equal to 1/24th of an ounce.

31. **Qian**: A Chinese unit of weight, used for silver and gold, equal to about 3.75 grams.

32. **Noggin**: An informal unit of volume for liquids, equal to about 1/4th of a pint.

33. **Hektoen**: A unit of measure for concentration in clinical chemistry.

34. **Quire**: A unit of paper quantity, typically 24 or 25 sheets.

35. **Morgen**: A unit of land area in various European countries, roughly equivalent to an acre.

36. **Scandal**: A unit of measurement for assessing the intensity of political scandals.

37. **Finger**: An informal unit of length, typically the width of a finger.

38. **Dunam**: A unit of land area in the Middle East, equal to 1,000 square meters.

39. **Rood**: An old English unit of land area, equal to a quarter of an acre.

40. **Cubit (Royal)**: An ancient Egyptian unit of length based on the Pharaoh's forearm.

41. **Dekagram**: A unit of mass equal to 10 grams.

42. **Qindarka**: An Albanian unit of currency, no longer in use.

43. **Cloves**: A unit of quantity for spices, especially cloves, often used in medieval trade.

44. **Bee's Knees**: An expression used to describe something exceptionally small or insignificant.

45. **Bale (Cotton)**: A unit of weight for cotton, typically weighing 500 pounds.

46. **Candlepower**: A unit of luminous intensity, especially used for measuring the brightness of lamps.

47. **Candy**: A unit of mass used in Japan, equal to 1000 grams or 1 kilogram.

48. **Darwin (Evolutionary)**: A playful unit used to describe the rate of evolution.

49. **Firkin**: A unit of volume for liquids, typically used for beer, equal to about 9 gallons.

50. **Olympic Swimming Pools**: A fun unit used to measure large volumes, equal to the volume of water in an Olympic-sized swimming pool.

51. **Mite**: A tiny unit of currency, often used in biblical contexts.

52. **Attogram**: A unit of mass equal to one quintillionth (10^{-18}) of a gram.

53. **Pony**: A small unit of volume for alcoholic beverages, typically half a jigger or one ounce.

54. **Sagan**: A unit of measurement used humorously to describe large quantities, named after astrophysicist Carl Sagan.

55. **Quadrant (Astrological)**: A unit of a celestial circle, typically measuring 90 degrees.

56. **Nail (Length)**: An ancient unit of length used for cloth, equal to 1/16th of a yard.

57. **Slug**: A unit of mass in the Imperial system, equal to 32.174 pounds-force·second^2/foot.

58. **Quack (Sound)**: A humorous unit of sound used to describe the sound produced by a duck's quack.

59. **Jeroboam**: A unit of volume for wine or champagne, often equivalent to 3 liters.

60. **Blink (Time)**: An informal unit of time, roughly equal to the time it takes to blink.

61. **Wien (Frequency)**: A unit of frequency used in physics, equal to 3.28984 x 10^15 hertz.

62. **Barn Megaparsec**: A humorous unit used in astrophysics to describe the size of certain cosmic structures.

63. **Slug (Speed)**: A playful unit of speed used to describe the pace of a slow-moving object.

64. **Smoot-Hawley Tariff**: A humorous unit of economic impact, referring to the effects of a historic tariff.

65. **Drake (Radio)**: A unit used in radio astronomy to measure radio source flux density.

66. **Femtoacre**: A humorous unit of land area, equal to 10^-15 acres.

67. **Teaspoon (Angle)**: An informal unit of angle, typically used to describe a small angle.

68. **Helen (Face Value)**: A playful unit of beauty used to describe an exceptionally beautiful face.

69. **Boltzmann's Constant**: A fundamental constant used in physics to relate energy to temperature.

70. **Carat (Gemstone)**: A unit of weight for gemstones, equal to 200 milligrams.

71. **Sverdrup (Spicy)**: A humorous unit used to describe the spiciness of food.

72. **Bloop (Ocean)**: A unit of sound used in underwater acoustics to describe certain mysterious underwater sounds.

73. **Lick (Astronomical)**: A unit of distance used in astronomy, equal to 3.09 x 10^16 kilometers.

74. **Cricket Chirp (Temperature)**: A playful unit of temperature used to estimate temperature based on cricket chirping.

75. **Bark (Sound)**: A unit of measurement for the loudness of sounds, especially used in the field of psychoacoustics.

76. **Bam (Impact)**: A humorous unit of impact, often used to describe a sudden and significant impact.

77. **Knot (Speed)**: A unit of speed used in navigation, equal to one nautical mile per hour.

78. **Bolt (Lightning)**: A playful unit used to describe the speed of lightning.

79. **Planck Length**: The smallest possible length in the universe, approximately 1.616×10^{-35} meters.

80. **Nautical Mile**: A unit of distance used in navigation, equal to one minute of latitude on the Earth's surface.

81. **Newton (Apple)**: A playful unit used to describe the force required to accelerate an apple at 1 meter per second squared.

82. **Cord (Wood)**: A unit of volume for firewood, equal to 128 cubic feet.

83. **Pennyweight (Precious Metals)**: A unit of weight for precious metals, equal to 1/20th of a troy ounce.

84. **MilliHelen (Beauty)**: A humorous unit of beauty, referring to the amount of beauty required to launch one ship.

85. **Liberty (Statue)**: A playful unit of height, often used to describe the height of statues.

86. **Parsec**: A unit of astronomical distance, approximately 3.086×10^{16} meters.

87. **Megaparsec**: A unit of astronomical distance, equal to one million parsecs.

88. **Pferdestärke (PS)**: A unit of power used in Germany, equivalent to approximately 0.98632 horsepower.

89. **Candela**: A unit of luminous intensity, used in the International System of Units (SI).

90. **Wolfram**: A unit of mass used in nuclear physics, equal to one atomic mass unit (approximately 1.66054×10^{-27} kilograms).

91. **Sone**: A unit of loudness used in acoustics, quantifying the perceived loudness of a sound.

92. **Bee Space (Beekeeping)**: A space in a beehive that is just wide enough for bees to move freely.

93. **Quintal (Metric)**: A unit of mass equal to 100 kilograms, used in some countries for agricultural products.

94. **Pluto Year**: The time it takes for Pluto to complete one orbit around the Sun, approximately 248 Earth years.

95. **Hand (Horse Height)**: A unit of horse height, equal to 4 inches or 10.16 centimeters.

96. **Sack (Coffee)**: A unit of coffee weight used in trade, equal to 132 pounds or 60 kilograms.

97. **Elephant (Prints)**: A humorous unit used to describe large and easily noticeable footprints.

98. **Dol (Sound)**: A unit of measurement for the perceived loudness of sounds, often used in architectural acoustics.

99. **Sheep (Power)**: A playful unit of power used to describe the output of a sheep (about 0.1 horsepower).

100. **Donkey (Distance)**: A whimsical unit used to describe a short distance, roughly the length a donkey might travel.

www.ingramcontent.com/pod-product-compliance
Lightning Source LLC
Chambersburg PA
CBHW072228290526
45794CB00007B/2938